Energ
235

# 扑灭森林火灾

## Fighting Forest Fires

### Gunter Pauli

[比] 冈特·鲍利 著

[哥伦] 凯瑟琳娜·巴赫 绘

贾龙智子 译

上海远东出版社

# 丛书编委会

主　任：贾　峰

副主任：何家振　闫世东　郑立明

委　员：李原原　祝真旭　牛玲娟　梁雅丽　任泽林

　　　　王　岢　陈　卫　郑循如　吴建民　彭　勇

　　　　王梦雨　戴　虹　靳增江　孟　蝶　崔晓晓

特别感谢以下热心人士对童书工作的支持：

匡志强　方　芳　宋小华　解　东　厉　云　李　婧

刘　丹　熊彩虹　罗淑怡　旷　婉　杨　荣　刘学振

何圣霖　王必斗　潘林平　熊志强　廖清州　谭燕宁

王　征　白　纯　张林霞　寿颖慧　罗　佳　傅　俊

胡海朋　白永喆　韦小宏　李　杰　欧　亮

# 目录

# Contents

地球正变得越来越暖和。到处都是森林火灾。有一只鹿和一只沙袋鼠，他们想知道自己能做点什么来减少发生火灾的风险。

　　"看看我们鹿在这里是多么勤勤恳恳地吃草。我们通过每天吃这些草和灌木，尽自己的力量来阻止森林火灾的发生。"鹿说道。

The Earth is getting warmer and warmer. There are wildfires everywhere. A deer and a wallaby wonder what they can do to reduce the risk of fire.

"Look at how diligently we deer are grazing and browsing here. By eating all these grasses and shrubs every day, we are doing our bit to stop wildfires," the deer says.

······到处都是森林火灾······

... wildfires everywhere ...

我们吃了这么多落叶……

We eat so much leaf litter ...

"而我们沙袋鼠吃了这么多覆盖在森林树冠下面土地上的落叶，以至于几乎没留下什么可以引起火灾的燃料了。"

"我经常为了够到树木较低枝丫上的叶子而伸长自己的脖子。通过这种方式，我们鹿在低矮的树枝和林地干燥的灌木丛之间制造了一个空隙，形成完美的防火屏障。"

"And we wallabies, we eat so much leaf litter lying on the ground under the forest canopy, that there is hardly any fuel left for fires."

"I often stretch my neck to reach the leaves on the lower branches of trees. In this way, we deer make a gap between low tree branches and the dry undergrowth on the forest floor, forming a perfect fire break."

"你已经搞定了！我们所做的，是阻止小火点燃树木并烧到树冠，摧毁住在那里的一切生物。"

"不过，冠军消防员是长颈鹿。有了她，大草原上的草地野火才不会轻易蔓延到树上。长颈鹿对保护我们的安全至关重要。"

"You've got that covered! What we do, is to stop small fires from burning up the tree and reaching the canopy, destroying everything living there."

"The champion firefighter, though, is the giraffe. With her around, grass fires in the savannah will not easily spread to trees. Giraffe are so important to keep us safe."

冠军消防员是长颈鹿。

The champion firefighter is the giraffe.

通过每天走动清理赛道。

Clearing game trails just by walking everyday.

"是的，我们动物不仅在垂直方向上创造了防火墙，在水平方向上也建造了防火屏障。"

"而且我们都沿着相同的赛道行进，清除沿途的植被。虽然我们只是像平日做的那样四处走动，但也创造了另一种安全措施。"

"Yes, we animals are not only creating vertical fire breaks, we're also making horizontal barriers."

"And we all walk along the same game trails, clearing them of vegetation. That is creating another safety measure – just by walking around as we do everyday anyway."

"许多人可能会认为这些行为太微小，没什么用处。而我认为，这是一种简单但非常有效的能够阻止火势蔓延的方法。"

"而且别忘了，我们的小朋友蚂蚁也帮了忙。无论他们在哪里建立殖民地，他们小小的脚和微小的重量都会创造一些最好的早期防火屏障。"

"Many people may dismiss this as being too little to be of any use. I, for one, think it a simple, yet very effective, way to stop fires spreading."

"And don't forget our tiny friends, the ants, who also help. Wherever they establish a colony, their tiny feet and minuscule weight will create some of the best early fire barriers."

我们的小朋友蚂蚁……

Our tiny friends, the ants …

老鹰扔燃烧的木头抓老鼠。

Eagles dropping buring wood to catch rats.

"不幸的是，不是每个人都知道如何降低火灾风险。我怀疑有些动物甚至热衷于放火，这样他们就可以得到一顿免费的饭了……"

"没错。我见过老鹰到处扔燃烧的木头，去抓躲避火的老鼠。但也有其他动物并不知道他们的口腹之欲引起了火灾。"

"Unfortunately, not everyone knows how to reduce the risk of fire. I suspect some animals are even keen on starting fires, so that they can get a free meal…"

"True. I have seen eagles dropping burning wood around, to catch the rats fleeing the fire. But there are others, who have no idea that their appetite causes fire."

"你这是在指责什么人吗？"

"没有指责，只是在解释。你知道爱吃树叶的小网蝽吗？他们巨大的胃口会导致叶子开始变臭。但是，这些恶臭的叶子更容易被点燃。"

"我不明白……"

"Are you accusing anyone?"

"No accusations, only explanations. You know the little lace bugs that love to eat leaves? Their huge appetites will cause the leaves to start tasting foul. But those foul-tasting leaves will then burn easier."

"I do not get that…"

小网蝽……

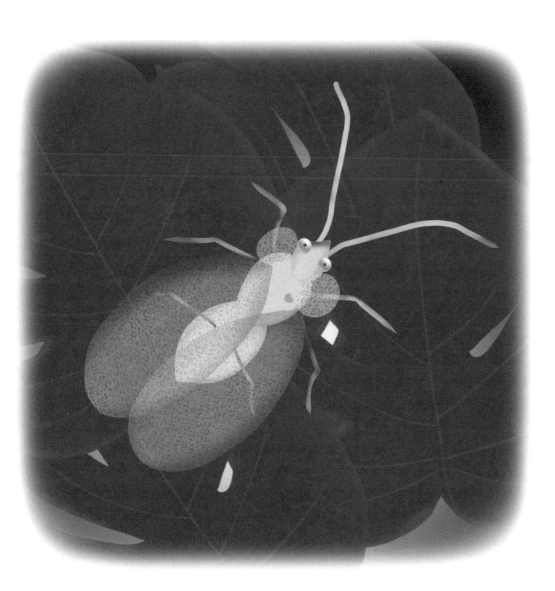

little lace bugs …

啮齿动物用树叶筑巢……

Rodents that build nests using leaves ...

"如果叶子不好吃，它们就不会被吃掉。这些叶子的降解速度会变慢，因此更多的燃料会积聚起来，给熊熊燃烧的火焰提供动力。"

"哇，你很聪明！你将所有点连接成了一个逻辑串，展示如何通过共同合作来保护每个人免受火灾之苦。"

"也有啮齿动物用树叶筑巢。他们把落叶堆积起来，就像篝火一样——随时可能燃烧！"

"If the leaves are not tasty they won't be eaten. They will degrade slower, and so more fuel builds up, to power a raging fire."

"Wow, you are smart! You connect all the dots in one logical string of thoughts, of how we can all work together to protect everyone from fire."

"There are also rodents that build nests using leaves. They pile them up, and that works like a bonfire – just ready to burn!"

"那么，我们该怎么办呢？"

"改变旧习惯是很困难的，但无论如何，每个人都必须学会适应。不再使用篝火和烟花。齐心协力，我们将做出改变，为了更好的未来！"

……这仅仅是开始！……

"So what do we do about that?"

"It is difficult to change old habits but somehow everyone will have to learn to adapt. Bonfires and fireworks are taboo. Working together, we will make changes, for the better!"

… AND IT HAS ONLY JUST BEGUN!…

……这仅仅是开始！……

... AND IT HAS ONLY JUST BEGUN! ...

The eating habits of goats, deer, and cows reduce the biomass available for potential wildfires. Birds, termites, and elephants naturally reduce the chances of wildfires spreading.

山羊、鹿和奶牛的饮食习惯减少了引起潜在森林火灾的生物量。鸟类、白蚁和大象的习惯自然而然地减少了野火蔓延的概率。

Grazers are effective fire managers in grassy habitats like savannah. Domesticated grazers can, however, promote the growth of less tasty plants, ones that are often drier and more flammable.

食草动物是草原等草木栖息地有效的火灾管理者。然而，驯养的食草动物会促进不那么美味的植物的生长，这些植物通常更干燥、更易燃。

Insects that feed on leaves, stimulate the production of defensive chemicals in those plants, changing the flammability of their leaves through the increase of lignin.

以叶子为食的昆虫，刺激这些植物产生防御性化学物质，通过增加木质素改变了叶子的可燃性。

Termites create massive structures that can house a variety of animals. Herbivores like to graze around the termite mounds, trampling the area and making it less likely to burn, creating bushfire safety zones.

白蚁创造出可以容纳多种动物的巨大结构。食草动物喜欢在白蚁丘周围吃草，践踏该区域，使其不太可能燃烧，从而形成丛林防火安全区。

Animals manage fire-spread by rearranging plants or dead leaves within their habitats. The mallee fowl gathers dead leaves into piles to incubate their eggs, helping to clear the ground of leaf litter.

动物通过重新安排栖息地内的植物或枯叶来控制火势蔓延。麻雀把枯叶堆成一堆，在其上孵化自己的蛋，帮助清理了地面上的落叶。

Animal tracks act like roads, creating firebreaks that can stop the advancing fire. Elephants trample plants, to form wide corridors running through dense foliage.

动物的足迹就像道路一样，创造出可以阻止火势蔓延的防火带。大象践踏植物，在茂密的树叶中形成宽阔的走廊。

Bark beetle attacks weaken and kill trees, reducing their moisture content and thus making them ignite more readily. Leaf-cutter ants remove the leaves that could fuel fires, in and around their nests and foraging trail.

树皮甲虫的攻击削弱并杀死树木，降低树木的含水量，从而使其更容易点燃。切叶蚁会清除它们的巢穴和觅食路径内以及周围可能会引发火灾的树叶。

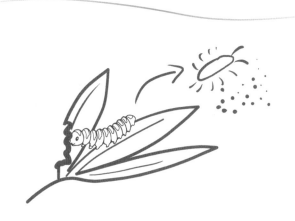

Invertebrates chew the dead leaves and twigs of the eucalyptus, breaking these into smaller pieces and making it easier for microbes to degrade it further, speeding up decomposition and reducing fire risk.

无脊椎动物会咀嚼桉树的枯叶和嫩枝，将它们分解成更小的碎片，使微生物更容易将其进一步降解，从而加快分解速度，降低火灾风险。

Are you aware of ways you may increase the risk of fire in a forest?

你是否意识到某些做法可能增加森林火灾的风险?

Do you think an ant colony reduces the risk of fire?

你认为蚁群能减少火灾的危险吗?

Why would a bird spread fire?

鸟为什么要传播火灾?

Is fire always bad for Nature?

火总是对大自然有害吗?

Fire poses a great risk to all life. Forest fires are intensifying due to climate change, and as it gets warmer and warmer, forests tend to get drier and drier. Start a research project with your friends, and family members, looking into ways in which we can reduce the risk of forest fires. Have each team draw up two lists, one of things we should definitely do, and another of the things we should definitely not do. Compare your lists and draw up a master list of the most important things we can do to reduce fire risk.

火灾对所有生命都构成极大的危险。由于气候变暖，森林变得干燥，发生森林火灾的风险正在加剧。与你的朋友和家人一起，研究降低森林火灾风险的方法。让每个小组列出两个清单，一个是我们绝对应该做的事情，另一个是我们绝对不应该做的事情。比较一下你们的清单，并列出我们为降低火灾风险而能做的最重要的事情。

## 学科知识
### Academic Knowledge

| | |
|---|---|
| **生物学** | 海滩松及其近亲短叶松的球果只有被火点燃时才会释放种子；血液中一氧化碳含量高的迹象包括头痛、头晕、恶心和精神功能减退；沙袋鼠和袋鼠的区别；草和灌木之间的区别；鸟类助长火势蔓延。 |
| **化 学** | 木质素易燃；森林火灾的烟雾中含有一氧化碳，这是一种无色、无味的有毒气体；阻燃剂的化学性质；当花边虫攻击树叶时，植物如何产生芥子油；燃烧需要氧气。 |
| **物 理** | 每天击中地球表面的闪电超过10万次，只有极少数确实引起了火灾；森林火灾上坡速度比下坡快；坡度越陡，火势传播速度越快；森林地面上的火苗高度可能达到1米，温度可达800摄氏度。 |
| **工程学** | 在森林火灾风险较高的地区，控制性燃烧是常见做法；使用飞行器（灭火飞机）进行空中灭火；协调操控无人机进行灯光表演正在取代烟花爆竹。 |
| **经济学** | 森林火灾期间的工资和就业增加；大规模森林火灾造成的短期劳动中断，由于需要扑灭工作而得到平衡。 |
| **伦理学** | 全球变暖导致了更多的森林火灾，而控制碳排放的努力却很少；大多数森林火灾都是人为的，要么是用火不慎，要么是森林管理不当。 |
| **历 史** | 最近的森林火灾强度和频率在过去1万年来是最为严重的；以湖泊沉积物中保存的木炭和花粉为基础的古生态研究，让我们了解了几千年来的森林火灾。 |
| **地 理** | 全新世的森林火灾频繁发生，造成显著的侵蚀和山坡退化；比起火灾后未发生降雨事件的燃烧区，火灾后的风暴会产生更大的侵蚀率；热带雨林通常很少发生森林火灾，因为潮湿的大气会充当巨大的防火屏障。 |
| **数 学** | 3D操作，以及垂直和水平方向操作；数百万计的微小行动与几个非常大的行动的对比；特定生物群落中的扰动可变性可用来计算火灾风险度。 |
| **生活方式** | 被认为是"消防安全"的家庭应该在30米范围内没有易燃植物或可燃材料。 |
| **社会学** | 指责与寻求解释之间的区别；篝火是一种作为庆祝活动组成部分的可控的户外火。 |
| **心理学** | 纵火癖是一种罕见的病理障碍，其特点是故意和反复纵火；纵火癖患者对火非常着迷，一旦纵火成功，就会体验到满足感或释放内心的紧张或焦虑。 |
| **系统论** | 大型森林火灾或火灾通常能够改变当地的天气状况或产生"自己独特的天气系统"；破坏性的火灾烧毁了原本可以防止侵蚀的植物和树木；火灾后发生的暴雨导致山体滑坡、灰烬流和山洪暴发，影响溪流、河流和湖泊的水质；许多因素集合在一起可能引起风险，也可能降低风险。 |

## 情感智慧
### Emotional Intelligence

**鹿**

鹿对自己在进食的同时做了对环境有益的事感到自豪，但她也承认其他动物，比如长颈鹿，在抑制火势方面更有成效。她洞察敏锐，对所有参与者的看法是平衡且互相联系的，包括那些造成负面影响而不自知者。当沙袋鼠向鹿祝贺时，她仍然非常谦虚，甚至没有回应那些赞美之词。她分享了她对他人所做出的正面或负面贡献的广泛的观点，此外，她还提供了一个具有前瞻性的观点。

**沙袋鼠**

沙袋鼠知道自己在降低火灾风险方面的作用。他的行为有助于创造空间来预防火灾。他认可森林中较小物种做出的贡献。他指出了一些动物忽视了火灾，还有另一些动物急于利用火为自己谋利。他担心鹿在暗指自己不知道导致森林火灾蔓延的原因。他勇于询问鹿是否在指责他自己。虽然他并不理解鹿提出的所有论点，但她广博的知识给他留下了深刻的印象。沙袋鼠称赞了鹿聪明的做法，以示对鹿的共鸣。

## 艺术
### The Arts

沙袋鼠是受欢迎的卡通形象，为什么不画出你最喜欢的沙袋鼠卡通形象呢？好好观察沙袋鼠，研究它最显著的特征。你的画不必复杂，即使是一个简单的线条画也能突出两三个最突出的特征：嘴、尾巴和育儿袋。尽情地突出这些特点，然后看看你的朋友们是怎么注意到这些显著特征的。

## 思维拓展
### Systems: Making the Connections

气候变化影响森林火灾的规模和频率，我们该着手研究动植物在减轻森林火灾中扮演的角色了。生态系统的演变模式为我们提供了一个研究框架来设想这些动植物会如何演变，以适应火灾和流行病肆虐的新的生活环境。

有些动物吃草和觅食的习性对森林火灾起到了干预作用。山羊在干燥灌木丛里觅食降低了火灾风险。长颈鹿从3米高的金合欢树上摘取叶子，有助于防止灌木丛的火蔓延到树冠。大象成群的活动创造了宽阔的林间走廊。许多小型动物，如野兔以及白蚁也做出了自己的贡献。它们制造的通道在火灾初始阶段起到了很好的隔离效果，保护其他野生动物免于灭顶之灾。

生态系统有提高自身恢复力的天然方式。有些植物需要火让种子发芽，而有些鸟类则传播火来引出躲起来的啮齿动物。总的来说，生态系统通过动植物间的相互作用规避风险。真菌也通过保持着枯木中的水分为遏制火灾做出贡献。大自然的策略与人类的方法有着本质不同。我们过度使用难以降解的化学物质作为阻燃剂，还会对健康造成危害。我们应该看到自然界的贡献，并为所有生物营造更安全的环境。

## 动手能力
### Capacity to Implement

森林火灾不仅对动植物很危险，对人类也很危险。我们需要摆脱对阻燃剂的依赖，或许应该开始更多地思考动植物应对火灾风险的方式。起草一份清单，列出你可以在家里或周围、花园、学校和当地公园里做的事情，这些事情的灵感来源于动植物应对火灾风险的聪明方法。这样做的好处是，它不仅会让你停止做对环境有害的事情，而且会让你乐于做对环境以及所有身处其中的动物和人类有益的事。

## 故事灵感来自
## This Fable Is Inspired by

# 克莱尔·福斯特
# Claire Foster

克莱尔·福斯特于 2008 年获得西澳大利亚大学理学学士学位。毕业后，她在澳大利亚野生动物保护协会工作，调查澳大利亚中部的动植物。2015 年，她在澳大利亚国立大学获得博士学位。她的博士研究调查了计划烧除和食草行为对桉树林下层植被的动植物群的交互作用。她现在是澳大利亚芬纳环境与社会学院的研究员。她研究动物群、燃料和火灾之间的相互作用，重点研究本土动物对燃料动力和丛林火灾风险的影响。

**图书在版编目（CIP）数据**

冈特生态童书.第七辑：全36册：汉英对照 /
（比）冈特·鲍利著；（哥伦）凯瑟琳娜·巴赫绘；
何家振等译.—上海：上海远东出版社，2020
ISBN 978-7-5476-1671-0

Ⅰ.①冈… Ⅱ.①冈… ②凯… ③何… Ⅲ.①生态
环境–环境保护–儿童读物—汉英 Ⅳ.①X171.1-49

中国版本图书馆CIP数据核字（2020）第236911号

**策　　划**　张　蓉
**责任编辑**　程云琦
**助理编辑**　刘思敏
**封面设计**　魏　来　李　廉

冈特生态童书

**扑灭森林火灾**

[比]冈特·鲍利　著
[哥伦]凯瑟琳娜·巴赫　绘

贾龙智子　译

记得要和身边的小朋友分享环保知识哦！
八喜冰淇淋祝你成为环保小使者！